How to Calculate?

I0468218

The nth root

of a positive real number

(Saini's nth root method)

By

Babu Lal Saini

Book availability

Hard Copy: *On line stores like--Amazon. in, Marketplace,*
Amazon.com, Flip kart, Pothi.com, Infibeam

Soft Copy: *All Amazon Stores*

Saini's nth root method

USD: 1.5 only

Saini's n^{th} root method

Preface

At present there does not exist a long hand method for finding the higher roots (say 3rd , 4th , 5th etc.) of a positive real number.

The formula suggested here, enables us to fid any root of a positive real number. With or without decimal.

Babu Lal Saini

S-96, Shivalik Nagar
BHEL, Haridwar
(Uttarakhand)
email: sainibl96@gmail.com
ph.   +919897982154

Saini's formula for n^{th} root of a positive real number

The method for calculating the nth root of a positive realnumber is as follows-

Step- 1

Make the groupsof n digits each. The rules for this are –

A. For real numbers without decimals-

Start from right side making the groups of n digits each. The last group may be left with a lesser no. of digits, if the no. of total digits is not fully divisible by n.

e.g.- $12\overline{34}5$

B. For real numbers with decimals-

The grouping will start from the decimal, i.e. from right to left for the digits before decimal and from left to right after decimal.

e.g.-

$$\overline{123}\overline{45}.\ \overline{123}\overline{45}$$

Step -2

Draw one line above the number and another on its left side as shown in Fig. 1.

Fig. 1

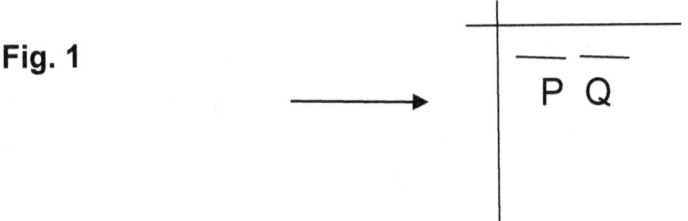

Step- 3

Consider the first group and Choose the highest real number **'a'** such that a^n is equal to or less than the first group.

Step -4

Write 'a' above the line and a^{n-1} on the left side. The product of a and a^{n-1} i.e. a^n will be written below the first group as shown in the figure- 2

Step -5

Take the difference of the two numbers(i.e. the first group and a^n. Now put the second group on the right side of the difference obtained. If it does not make a sufficient big number, put the third group on the right side of it and one zero after a. In the example shown below, this number is $(P-a^n)Q = C$ (say).

Fig.- 2 ⟶

$$
\begin{array}{c|l}
 & a\ . \\
\hline
a^{n-1} & \overline{P}\ \overline{Q}\ \overline{R} \\
 & -a^n \\
\hline
 & (P-a^n)Q
\end{array}
$$

Step-6

Now write, in the left corner,

$$\frac{n}{1!} \times 10^{n-1} \times a^{n-1} + \frac{n(n-1)}{2!} \times 10^{n-2} \times a^{n-2} \times (- -)^1 +$$

$$\frac{n(n-1)(n-2)}{3!} \times 10^{n-3} \times a^{n-3} \times (- -)^2$$

$$+ \frac{n(n-1)(n-2)(n-3)}{4!} \times 10^{n-4} \times a^{n-4} \times (- -)^3 \text{-----}$$

$$up\ to \quad a \times (- -)^{n-2} \quad + (- -)^{n-1}$$

The blank spaces are to be filled by some digit **b** such that the final number obtained is equal or less than **c/b**. The digit **b** should be decided by trial and error so that its highest value is taken.

The number now becomes-

$$\frac{n}{1!} \times 10^{n-1} \times a^{n-1} + \frac{n(n-1)}{2!} \times 10^{n-2} \times a^{n-2} \times b^1 +$$

$$\frac{n(n-1)(n-2)}{3!} \times 10^{n-3} \times a^{n-3} \times b^2$$

$$+ \frac{n(n-1)(n-2)(n-3)}{4!} \times 10^{n-4} \times a^{n-4} \times b^3 \text{---------}$$

$$up\ to \quad a \times b^{n-2} \quad + b^{n-1}$$

*Let us call this number as **Saini number (Sn)***

Step – 7

Put the Saini number in the bottom left corner as shownin the figure-3.

$$
\begin{array}{c|ccc}
 & a\,. & & \\
\hline
a^{n-1} & \overline{P} & \overline{Q} & \overline{R} \\
 & -a^{n} & & \\
\hline
Sn & (P-a^{n})Q & & \\
\end{array}
$$

Fig.-3 $\longrightarrow$

Note- *1. The Saini number, when calculated, is not so large as it appears to be. As it will actually have n-1 steps only. e.g. when we calculate 4th root, the Saini number will have only 3 steps i.e.-*

$4 \times 1000 \times a^3 + \dfrac{4 \times 3 \times 100}{2}\,a^2 \times b$

$+ \dfrac{4 \times 3 \times 2 \times 10 \, a \times b^2}{6} + b^3$

2. The digit b is decided in such a way that (S.N.)x b is less than or equal to the number $(P- a^{n})Q$

Step- 8

Put **b** above *the top line , i.e. after* **a.**

Step-9

Multiply Sn by **b** and put it below $(P-a^n)Q$.

Step -10

Subtract $(Sn)*b$ from $(P-a^n)Q$ and get the remainder.

Step-11

If some unused group (of n digits) is still left, put it after the remainder to make a fresh number. Repeat steps from 3 to 10 with a difference that figure **ab** will be taken in place of *a* for calculating the Sn.

Step-12

If the remainder is zero, the process is complete. The root calculated is **ab**.

But, If the remainder (say d) is not zero, than the root is not an integer and we will have to calculate further i.e. after decimal.

Step-13

For this purpose, put a decimal after **ab**, on the top line and put n zeros after the remainder (d).

Step-14

Repeat the steps from 3 to 10 taking **ab** in place of a for calculating the Sn.

Hence,

for cube root the Sani number (Sn) will be-

$Sn_3 = 300a^2 + 30\ ab + b^2$

Similarly for 4th root-

$Sn_4 = 4000\ a^3 + 600\ a^2b + 40\ ab^2 + b^3$

And for 5th root-

$Sn_5 = 50000a^4 + 10000a^3b + 1000a^2b^2 + 50\ ab^3 + b^4$

Some examples to elaborate the method are given ahead-

Example – 1

To calculate the 3rd root of 2

	$1(a).2(b)5(b_1)$
$1(a^{2)}$	$\overline{2.000}\ \overline{000}$
	$1(a^{3)}$
	$-$
$300+30*2(b)+ 4(b^2)$ $=364(c)$	1000 $728\ (c*b)$
	$-$
$300*12^2 +30*12*b_1 +$ $(b_1)^{\underline{2}} =$ $43200+360*5(b_1)+$ $25(b_1{}^2)= 45045\ (Sn_1)$	272000 $225125\ (Sn_1*b_1)$ $-\!\!-\ \overline{}$
	4875

So, the 3rd root of 2 is 1.25…..

Explanation:

Here, the original number is 2, so we put 3 zeros after decimal and make groups as shown above.

So, the first group will be 2.

Hence, a =1,

we put 1 on the left side of the vertical line and 1 below 2 and subtract. The remainder is 1.0 .

Now, we take the second group, i.e. 000. So, the new figure becomes 1000.

Since, we have taken a group after decimal, so we put a decimal after 1(a) above the top line.

Here the Saini number (Sn) will be—

Sn= **300 + 30*b +b^2**

taking b=1,

Sn = 300+30+ 1 = 331

So, Sn*b=331*1=331

Which is much less than 1000

Next, we take b=2.

So, Sn=

$300+30*2+2^2 == 364$

And Sn*b= 364*2= 728, which is close to 1000.

Now, put 2 i.e. b, after decimal above the top line and put 728, below 1000 and subtract.

The remainder is 272

Now, we take the third group i.e. 000 and put it after 272

Hence, the new figure is 272000.

Now, we calculate Sn for 272000

(Here, a_1 =12)

So, new Sn -

$Sn_1= 300*12^2 +30*12*b_1 +b_1^2$

By hit and trial we find that optimum value of b_1 is 5.

So,

$Sn_1 = 300*12^2 + 30*12*5 + 5^2 = 45045$

And $Sn_1 *b = 45045*5 = 225125$.

Now, put 5 i.e. b_1 after 2, above the top line and 225125 below 27000 and subtract.

The remainder is 4875.

.This process can be continued further. But to stop it here only, we conclude that-

The 3rd root of 2 is 1.25…..

Example – 2

To calculate the 4th root of 2

	$1(a).1(b)8(b_1)$
$1(a^3)$	$2.\overline{0000}$
	$1(a^4)$
	$-$
$4000+600*(b)+$ $40(b^2)+b^3=4641(Sn)$	10000 $4641\ (Sn*b)$
	$-$
$4000*11^3+600*11^2*b_1 +$ $40*11*(b_1)^2+b_1{}^3 =$ $5324000+ 72600*(b_1)+$ $40*b_1{}^2+b_1{}^3$ $=5933472(Sn_1)$	53590000 47467778
	$-$
	6122222

So, the 4th root of 2 is 1.18…

Explanation:

Here also, the original number is 2.0000..

Now, a is to be decided in such a way that a^4 is less than or equal to 2. Hence, a shall be 1.

We put 1 i.e. a^3 on the left side of the vertical line and 1 i.e. a^4 below 2 and subtract.

The remainder is 1. So we put the next group i.e. 0000 Next to it. The new figure becomes 10000. Since a group after decimal has been taken, we put a decimal after 1 above the top line..

Now, Sn = **$4000\,a^3 + 600\,a^2b + 40\,ab^2 + b^3$**

Here, a=1, and taking b=1

Sn.=4000 + 600 + 40 +1= 4641

So, b x c= 4641

Now, we put this figure 4641 below 10000 and subtract. The remainder is 5359.

Now we take another group '0000' and put it after 5359. So, the number becomes 53590000.

Again we calculate Sn_1 taking $a_1 = 11$

So, $Sn_1 =$

$4000*11^3+600*11^2*b_1 +40*11*(b_1)^2+b_1^3 =$

$5324000+72600*(b_1)+40*b_1^2 +b_1^3$

By hit and trial, we find that $b_1 = 8$.

So, $Sn_1 = 5324000+72600*8+40*8^2 +8^3$
$= 5933472$

And , $Sn1*b_1 = 5933472*8 = 47467778$

Put this number below 53590000 and subtract.

The remainder is 6122222.

This process can be continued further.

To stop it here itself, we conclude that-

The 4th root of 2 is 1.18...

Example – 3

To calculate the 5th root of 2

	$1(a).1(b)4(b_1)$
$1(a^4)$	$\overline{2.00000}$
	$1(a^5)$
	$-$
$50000+10000*(b)+$ $1000*(b^2)+50b^3+b^4=$ $61051Sn)$	100000 61051 (Sn*b)
	$-$
$50000*11^4+10000*11^3*b_1$ $+1000*11^2*b_1{}^2+50*11*b13$ $+b_1{}^4 = 78722600$ (Sn)	3894900000 3148904000
	$-$
	745996000

So, the 5th root of 2 is 1.14…

Explanation:

Here also, the original number is 2.00000..

So, **a** has to be 1.

We put 1 i.e. a^4 on the left side of the vertical line and 1 i.e. a^5 below 2 and subtract.

The remainder is 1. So we put the next group i.e. 00000 Next to it, The new figure now is 100000. Since a group after decimal has been taken, we put a decimal after 1 above the top line..

So,
Sn = **$50000a^4 + 1000a^3b + 600a^2b^2 + 50\ ab^3 + b^4$**

Now, we decide **b** in such a way that its (Sn)xb is less than or equal to 100000.

Here, a=1, and taking b=1

Sn.=50000+1000+600+50 +1 =61051

So, Sn* b = 61051*1 = 61051

We put this figure 61051 below 100000 and subtract.

The remainder is 38949

Now we take another group '00000' and put it after 38949

So, the number becomes 3894900000.

Again we calculate Sn_1 taking $a_1 = 11$

So, $Sn_1 =$

$50000*11^4+10000*11^3*b_1+1000*11^2*(b_1)^2+50*11*b_1^3 +b_1^4 =$

By hit and trial, we find that $b_1 = 4$.

So, $Sn_1 = 78722600$

After multiplying this number with b_1 i.e. 4

We get $78722600*4 = 3148904000$

Put this number below 3894900000 and subtract. The remainder is 745996000

This process can be continued further.

To stop it here itself, we conclude that-

The 5th root of 2 is 1.14...

Example –4

To calculate the 3rd root of 1331

	1(a)1(b)
$1(a^{2)}$	$\overline{1331}$
	$1(a^{3)}$
	−
$300+30b+$ $b^2=331(Sn)$	331 331
	−
	0

So, 3rd root of 1331 = 11

Explanation:

Here, the original number is 1331, so we make the group as shown above.

Here, the first group will be 1.

Hence, $a = 1$,

we put 1 on the left side of the vertical line and 1 below 1 and subtract. The remainder is .0 .

Now, we take the second group, i.e. 331.

Taking $b = 1$, we calculate the Sn

$Sn = 300 + 30*b + b^2 = 331$

And $Sn*b = 331*1 = 331$

Now, we put 331 below 331 and subtract.

The remainder is 0.

Hence, we conclude that-

The 3rd root of 1331 is 11.

Example –5

To calculate the 3rd root of 10.2

	$2(a).1(b)\ 6(b_1)$
$4(a^{2)}$	$10.\overline{200}$
	$8(a^{3)}$
	—
$300*4+30*2*b+b2$ $= 1261(Sn)$	2200 1261
	—
$300*441+30*21*b_1+$ $b_1{}^2 =$ $132300+630b_1+b_1{}^2=$ 136116	939000 816696
	—
	122304

So, 3rd root of 10.2 = 2.16

Explanation:

Here, the original number is 10.2, so we put 2 zeros after decimal and make group as shown above.

Here, the first group will be 10.

Hence, a =2,

we put 2 (a) on the left side of the vertical line; 4 (a^2) on left side of the line and 8(a^3) below 10 and subtract. The remainder is 2. .

Now, we take the second group, i.e.200. So, the new figure becomes 2200.

Since, we have taken a group after decimal, so we put a decimal after 2(a) above the top line.

So, Sn= 300*2^2 +300*2*b+b^2

Taking b=1,

Sn. = 1200+600+1 = 1261

We put this figure, 1261 below 2200 and subtract.

The remainder is 939.

Now take another group '000' and put it after 939.

So, the figure becomes 939000.

Again we calculate Sn using $a_1 = 21$

$Sn_1 = 300 \cdot 21^2 + 30 \cdot 21 \cdot b_1 + b_1^2$

$\quad = 132300 + 630b_1 + b_1^2$

Now, by hit and trial we find that optimum value of b_1 will be 6.

So, $Sn_1 = 136116$

And $Sn_1 \cdot b_1 = 136116 \cdot 6 = 816696$

We put this figure below 939000 and subtract.

The remainder is 122304.

This process can be continued further. But to stop it here only, we conclude that-

The 3rd root of 10.2 = 2.16

Example –6

To calculate cube root of 1349.232625

$$1\,1.05$$

$1(a^2)$	$1\ \overline{349}.\overline{232}\ \overline{625}$
	$-1(a^3)$

$300 + 30 + 1 = 331(c)$

349
$-331(b_1c)$

$3630000 + 16500 + 25 = 3646525(b_1)$

$18232\ 625$
$-18232625(b_1c_1)$

x

So, $\sqrt[3]{1349.232\ 625} = 11.05$

Explanation:

Here, the original number is 1349.232 625. And, the groups of 3 digits are to be made (starting from right side before decimal and from left side after decimal).

Hence, the first group will be 1.

So, a =1

We put $1(a^2)$ on the left side of the vertical line and $1(a^3)$ below 1 of the total figure and subtract.

The remainder is zero.

Now, we take the second group, i.e. 349.

Now, Sn= . $300a^2 + 30\ ab + b^2$

Here a=1, taking b=1,

Sn. = 300 + 30 + 1 = 331

This is very close to 349.

So, b=1 is confirmed.

Now, put 1 i.e. b after 1 above the top line and again multiply 331 by b i.e. 1 and put it below 349 and subtract.

The remainder is 18.

Now, we take the third group i.e. 232 and put it after 18.

Hence, the new figure is 18232.

Since, this group 232 is after the decimal hence, place a decimal after 11 written above the top line.

Now, the new Sn i.e.

$Sn_1 = 300*a_1^2 + 30*a_1 * b_1 + b_1^2$

Here, $a_1 = 11$

So,

$Sn_1 = 300*11^2 + 30*11*b_1 + b_1^2$

Taking $b_1 = 1$

$Sn_1 = 36300 + 330 + 1 = 36631$.

This figure is more than the original one, i.e. 18232 and b_1 can not be less than 1, so we have to take a larger figure. For this purpose we put the last group i.e. 625 after 18232 to make the new figure as 18232625.

In lieu of bringing down the second group, we put a 0 i.e. zero above the top line as shown.

The new a i.e. a_1 is 110. (We don't consider the decimal during the calculations.)

So,

$Sn_1 = 300 * a_1{}^2 + 30 * a_1 * b_1 + b_1{}^2$

$\qquad = 300 * 110^2 + 30 * 110 * b_1 + b_1{}^2$

Taking $b_1 = 1$

$Sn_1 = 3630000 + 3300 + 1 = 3633301$

and $Sn_1 * b_1 = 3633301$
which is much less than 18232625.

So, we go for $b_1 = 2$

So,

$Sn_1 = 3630000 + 3300 * 2 + 2^2$

= 3630000 + 6600 + 4 = 3636604.

So, $Sn_1 * b_1$ = 3636604*2 = 7273208

This figure is also much less than 18232625.

Similarly,

for b=3, Sn_1 = 3630000 +9900 + 9 = 3639909,

So, $Sn_1 * 3$ = 3639909*3 = 10919727

and for b=4,

Sn_1 = 3630000 + 13200 + 16 = 3643216

and $Sn_1 * 4$ = 3643216*4 = 14568864

Lastly for b=5,

Sn_1 = 3630000 + 16500 + 25 = 3646525,

And $Sn_1 * 5$ = 364652*5 = 18232625,

which is equal to the desired number.

Hencethe answer is 11.05

Example – 7

To calculate 4th root of 234256

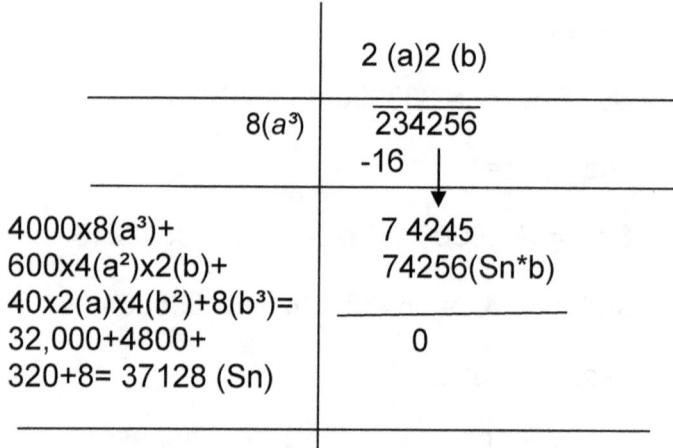

So, 4th root of 234256 = 22

Explanation:

Here, the original number is 23,4256. So, the groups of 4 digits are to be made (starting from right side before decimal and from left side after decimal).

Hence, the first group will be 23.

Now, **a** is to be decided in such a way that a^4 is less than or equal to 23. Hence, **a** has to be 2.

We put 8 i.e. a^3 on the left side of the vertical line and 16 i.e. a^4 below 23 and subtract.

The remainder is 7.

So we put the next group i.e. 4256 Next to it, The new figure now is 74256.

Further, we decide **b** in such a way that its Sn*b is less than or equal to 74256.

First, we take b=1

For this,

Sn.= 4000 a^3 + 600 a^2b + 40 ab^2 + b^3

= 4000x8 + 600x4 + 40x2 + 1) = 34481.

Which after multiplication by 1, remains 34481 only.

This figure   is much less than 74256.

So, we go for b=2.

New Sn.= 4000x8+600x8+40x8+8
        = 37128
And Sn*b = 37128*2 = 74256

Since, it is equal to the starting number, the remainder will be zero.

Hence, 4th root of 234256 = 22.

Example – 8

To calculate the 5th root of 234256

	2 (a)2 (b)
$16(a^4)$	$\overline{5}\,\overline{153632}$
	32
$50,000 \times 16(a^4)+$ $10,000 \times 8(a^3) \times 2(b)$ $+1000 \times 4(a^2) \times 4(b^2)$ $+50 \times 2(a) \times 8(b^3)+$ $16(b^4) =$ $800,000+16,0000$ $+16000+800+16=$ $976816 \,(Sn)$	19 53632 1953632 (Sn*b)
	X

So, 5th root of 5153632 = 22

Or, $\sqrt[5]{5153632} = 22$

Explanation:

Here, the original number is 5153632.

So, the groups of 5 digits are to be made (starting from right side).

Hence, the first group will be 51.

Now, a is to be decided in such a way that a^5 is less than or equal 51.

Hence, a has to be 2.

We put 16 i.e. a^4 on the left side of the vertical line and 32 i.e. a^5 below 51.

The remainder of 51-32 is 19.

So we put the next group i.e. 53632 next to it.

The new figure now is 1953632.

So,

$Sn = 5,0,000*2^4 + 10000*2^3*b + 1000*2^2*b^2 + 50*2*b^3 + b^4$

$= 80,0000 + 80000*b + 4000b^2 + 100b^3 + b^4$

Saini's n^{th} root method

Now, we decide **b** in such a way that Sn*b is less than or equal to 1953632.

First, we take b=1.

For this,

$Sn = 80,0000+80000*1+4000*1^2 +100*1^3+1^4$

$=80,0000+80,000+4000+100+1) = 884101$

which when multiplied by 1, remains 884100 only.

This figure is much less than 1953632.

So we go for b=2.

Now, new Sn.=

$= 80,0000 +10,000x16 + 1000x16 +100x8 +16 = 976816$

And Sn*b = 976816*2 = 1953632.

Since, it is equal to the starting number, the remainder will be zero.

Hence, 5th root of 5153632 = 22.

www.ingramcontent.com/pod-product-compliance
Lightning Source LLC
Chambersburg PA
CBHW070422190526
45169CB00003B/1375